Podcasting
Pocket Guide

GW00703301

Podcasting
Pocket Guide

Kirk McElhearn, Richard Giles,
and Jack D. Herrington

O'REILLY®

Beijing · Cambridge · Farnham · Köln · Paris · Sebastopol · Taipei · Tokyo

Podcasting Pocket Guide

by Kirk McElhearn, Richard Giles, and Jack D. Herrington

Copyright © 2006 O'Reilly Media, Inc. All rights reserved.
Printed in the United States of America.

Published by O'Reilly Media, Inc., 1005 Gravenstein Highway North,
Sebastopol, CA 95472.

O'Reilly books may be purchased for educational, business, or sales
promotional use. Online editions are also available for most titles
(*safari.oreilly.com*). For more information, contact our corporate/
institutional sales department: (800) 998-9938 or *corporate@oreilly.com*.

Editor:	John Neidhart
Production Editor:	Marlowe Shaeffer
Cover Designer:	Karen Montgomery
Interior Designer:	David Futato

Printing History:

October 2005:	First Edition.

0-596-10230-5
[C]

Contents

Introduction

Like this pocket guide, podcasts fit nicely in your pocket. Even if you use a computer to download them, you can still put them on something small so you can listen to your podcasts when and where you want. It's this *time-shifting* that leads some people to refer to podcasting as TiVo for radio. They are close, but they miss the most important thing: there is no radio!

Radio waves are a heavily regulated territory. As a result, you can't say what you want on them. In fact, most of you can't say *anything* on them, since you need deep pockets to run a radio station. Of course, there are exceptions: college radio stations are often very accessible to amateur DJs, and Low-Power FM (LPFM) licenses are available to certain organizations. But such opportunities are few and far between.

This is where podcasting comes in.

Anyone who can make an MP3 recording, host it on a web site, and publish a "feed" for it can be a podcaster. The next step from there is to get the recording listed in podcasting directories, and listeners all around the world will start finding it.

As a listener, podcasting's accessibility means that you'll have access to opinions and ideas that you wouldn't get on the radio. Certainly, this means that there will be a lot of noise to sift through. The beauty of it all is that the best podcasts get the most buzz, and the good stuff tends to bubble to the top.

You can get a lot out of podcasts if you choose the ones that are right for you: listen to unique views on current events, hear music you'd never hear otherwise, and enjoy sound-seeing tours that take you all over the world.

Who This Book Is For

This pocket guide will help you get started, both as a listener and as a podcaster. For the listener, you'll learn how to tune into podcasts and download them to your favorite portable device so you can listen to them when you want. You'll also get an overview of some of today's best podcasts. Because podcasters will promote other podcasts, you only need to start out with a few, and then start listening to new shows as you learn about them.

For the aspiring podcaster, you'll learn how to make and publish podcasts with the most basic of equipment: a computer, a microphone, and some free software. Once you've made that first recording, you can publish it online, get it listed, and start obsessing over the size of your audience!

Organization of This Book

This book is intended to guide you through getting started in the world of podcasting, both as a listener and as a creator of your own podcast.

Chapter 1 will take you through the process of using iTunes to find, subscribe, and listen to podcasts.

Chapters 2, 3, and 4 are a brief introduction to creating, recording, and editing your own podcast.

Chapter 5 is a selection of reviews of some of the most interesting and unusual podcasts available when this book went to press in October 2005.

For a detailed guide to creating your own podcast, we suggest *Podcasting Hacks* (O'Reilly, 2005).

How to Contact Us

We have tested and verified the information in this book to the best of our ability, but you may find that features have changed (or even that we have made mistakes!). As a reader of this book, you can help us to improve future editions by sending us your feedback. Please let us know about any errors, inaccuracies, bugs, misleading or confusing statements, and typos that you find anywhere in this book.

Please also let us know what we can do to make this book more useful to you. We take your comments seriously and will try to incorporate reasonable suggestions into future editions. You can write to us at:

O'Reilly Media, Inc.
1005 Gravenstein Highway North
Sebastopol, CA 95472
(800) 998-9938 (in the U.S. or Canada)
(707) 829-0515 (international/local)
(707) 829-0104 (fax)

To ask technical questions or to comment on the book, send email to:

bookquestions@oreilly.com

The web site for *Podcasting Pocket Guide* lists examples, errata, and plans for future editions. You can find this page at:

http://www.oreilly.com/catalog/podcastingpg

For more information about this book and others, see the O'Reilly web site:

http://www.oreilly.com

CHAPTER 1

Finding, Subscribing to, and Listening to Podcasts

As you've seen in the Introduction, podcasting, which was initially a blip on the radar screen of Internet content, came of age in 2005. In fact, we can tell you the exact date that podcasting hit the mainstream: Tuesday, June 28, 2005. How can we narrow down this date so precisely? It's the day that Apple released iTunes 4.9, the first version of the program that provided fully integrated podcast support.

iTunes was, of course, not the first program that let you find podcasts, subscribe to them, and download them. Podcasting initially took off because programs such as iPodder, Doppler, and Podcast Tuner were the first tools to work with the combination of audio files and RSS that are behind podcasting. But when Apple came into the game—which was almost natural, since part of the word "podcasting" comes from "iPod"—things became easy. So easy, in fact, that in this chapter we'll focus on using iTunes to show you how to find, subscribe to, download, and listen to podcasts. (See "Working with Other Podcast Software," later in this chapter, for more on other programs you can use to find and manage podcasts.)

Say Hey to iTunes

iTunes is the one-stop shop for podcasts. Just as the program is a versatile tool for ripping CDs, managing music, and syncing music to your iPod, it has all the features a podcast management program needs. It lets you do all of the following and makes everything you do with podcasting as simple as clicking a few buttons:

Finding podcasts

iTunes has a huge directory of podcasts. When iTunes first launched the directory, it contained over 3,000 podcasts, and dozens of new podcasts are added each week. You can search for podcasts by name, you can browse by category, and you can check out what others listen to by scanning the Top Podcasts, a list of podcasts with the most iTunes subscribers.

Subscribing to podcasts

After you find the podcasts you want to listen to, a couple of clicks adds a subscription to that podcast to your iTunes podcasts collection. You don't need to worry about URLs, file formats, or anything else. iTunes downloads the latest episode, then updates the podcast at regular intervals, downloading new episodes when they become available. You can also subscribe to podcasts that are not in the iTunes Podcast Directory by adding their URLs to a dialog.

Downloading podcasts

While iTunes automatically downloads podcasts when you subscribe, you can also download individual episodes, either to try out a new show or to get a hold of older shows from a program's archives. Again, it's just a click away.

Organizing and saving podcasts

Once you download a podcast, you can listen to it on your iPod, on another digital music player, or on your computer, and you can delete the podcasts you've already heard. But if you want to keep some of them, you can use iTunes to organize your personal archive of podcasts.

Listening to podcasts

You can use your iPod to listen to podcasts, of course, but you can also listen to them on your computer with iTunes. The program offers many interesting features, such as bookmarking, so you don't lose your place in podcasts you haven't finished listening to. Some podcasts have chapters, which let you skip to the parts you want to hear.

Syncing podcasts to your iPod

If you've got an iPod, you already know how easy it is to use iTunes to sync your music library to your portable device. Syncing podcasts is just as easy; iTunes gives you some useful options, letting you choose which podcasts you sync and how long you keep them on your iPod. It's easy to fill up an iPod with podcasts, so these features ensure that iTunes only copies the latest shows or the ones you want. If you don't have an iPod, you can copy podcasts to another portable music player—some players work with iTunes, and for others, you can manually copy files that iTunes has downloaded.

In the rest of this chapter, we'll tell you how to do all this and more. You'll be ready to find dozens of podcasts, or subscribe to many of the podcasts listed in the review section of this book, load them on your iPod, and start listening. Just make sure you have time to listen to all the podcasts you download!

Finding Podcasts

Thousands of people record podcasts and provide them for your listening pleasure. But how can you find the ones that will interest you? Well, you can use this book; in fact, if you're reading this book, you're probably interested in our selections and reviews. But you can also browse through the many podcasts available in the iTunes Podcast Directory, looking for new shows, browsing by category, and searching by keyword.

Apple intelligently combined not only tools for subscribing to and downloading podcasts but also a comprehensive directory of podcasts so you can search for shows that interest you. The iTunes Podcast Directory is housed in the iTunes Music Store—even though, as of press time, all the podcasts available are free. To get there, click the Podcasts icon in the iTunes source list (that's your podcast playlist, in a way, as you'll see soon), then click the arrow next to Podcast Directory at the bottom of this window, as shown in Figure 1-1.

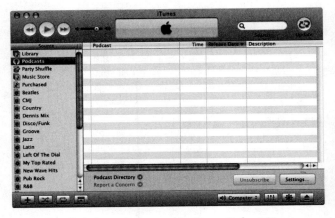

Figure 1-1. Click the arrow to go to the iTunes Podcast Directory.

This takes you to the main page of the Podcast Directory. Figure 1-2 shows this page, as it looked one day in July 2005.

If you're familiar with the iTunes Music Store, you'll find it easy to navigate this page and its links. If not, just think of it as a web page: the entire iTunes Music Store is designed like a web site, with graphic and text links to take you to other sections or to individual songs, albums, and programs.

Figure 1-2. The Podcast Directory in the iTunes Music Store is a great place to start browsing for podcasts.

If any of the featured programs—the ones with graphics or the ones in the Top Podcasts—interest you, click their links to see what they're about. Figure 1-3 shows an example of a page presenting a podcast.

Figure 1-3. If you're into science, here's what Science Friday offers.

Figure 1-3 shows the info page for the *Science Friday* podcast, a weekly show about science, hard and soft, featuring interviews and discussions with scientists and authors about the current hot topics in science. From this page, you can do the following:

Listen to a preview

Double-click a show to listen to a 1.5-minute preview of some shows. You can then decide if you want to hear more, either by subscribing or downloading an episode for further listening. For some shows, you can listen to an entire episode in this way.

Subscribe to a podcast

Click the Subscribe button to subscribe to this podcast. See "Subscribing to Podcasts," later in this chapter, for more on subscriptions and subscription options.

Download a single show

Click Get Episode to download a single show. See "Downloading Individual Shows," later in this chapter, for more information.

Visit the show's web site

Click Website to go to the show's web site. In some cases, you'll find additional information and resources. You may also find archives of shows that are not listed in the iTunes Music Store page or information on upcoming shows.

Complain

Click "Report a Concern" to fill out a form and send it to Apple if you think something is wrong with the show. For example, if it's explicit and should be rated as such, or if you think the show contains copyright violations.

Browsing the Podcast Directory

In addition to the podcasts featured on the main page of the Podcast Directory, you can browse through a few thousand other podcasts to find shows that might interest you. The breadth of podcast content, as you'll see in Chapter 5, is astounding, and the number of shows increases almost daily. Apple has sorted the podcasts available through iTunes by category—though, arguably, their category choices for some podcasts defy logic. To browse the Podcast Directory, go back to the main Podcasts page. In the lefthand column, click one of the categories in the list.

Clicking one of these links takes you into browse mode in the iTunes Music Store. You may be familiar with this type of columnar display if you've already browsed the Music Store; it's also very similar to the way you browse your iTunes music library. Figure 1-4 shows an example.

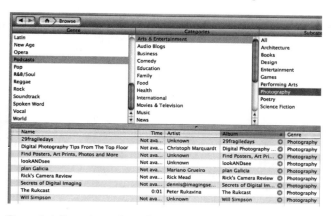

Figure 1-4. Browsing podcasts by category lets you see the full choice of programs you can subscribe to.

Some categories have subcategories, such as the Arts & Entertainment category shown in Figure 1-4, but others just lump together all the available shows. It's possible that, by the time this book goes to press, Apple will have created more subcategories to make browsing easier.

Searching for Podcasts

There's a third way you can find a podcast in iTunes: you can search for a keyword in its title, or for its creator—if you know the person's name. You can either enter a search string in the search field at the top of the iTunes window—if you are in the Podcast Directory, this shows that the search is for podcasts, as in Figure 1-2—or use the search field at the left of the window. You can search for either a keyword or "author," Apple's term for the show's creator, or you can limit your search to one of those search criteria by selecting it from the pop-up menu below the search field.

iTunes displays the results of your search with some of the top hits in the top section and the full results in the list at the bottom of the window, as shown in Figure 1-5. The search looks at show names and authors (or artists, as you see in Figure 1-5), as well as descriptions.

Figure 1-5. The results of a search for "gadget show".

If you find what you're looking for in these search results, you can do the following:

Go to a podcast's page

Click the arrow button next to the name of a podcast in the full results list to go to the program's page in the iTunes Music Store. Figure 1-6 shows the page for *The Gadget Show*.

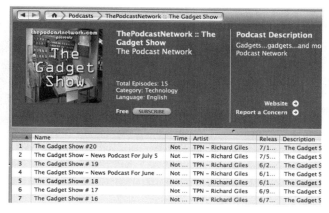

Figure 1-6. The iTunes Music Store page for Richard Giles' podcast, The Gadget Show.

Read a description of the podcast

Click the **i** icon in the Description field to display a window with a description of the show (see Figure 1-7).

Subscribe to the podcast

Click the Subscribe button to subscribe. See the next section, "Subscribing to Podcasts," for more on subscriptions.

Now that you've found the ear-candy you need, it's time to find out what you can do with your podcasts: how to subscribe and manage subscriptions, how to download podcasts, how to listen to them, and much more.

```
○ ○ ○                    Podcast Information
Podcast  - Skepticality - Science and Skeptic Thought
Episode  - Skepticality - Science and Skeptic Thought

We're tired of pertinent social and science news being buried
by clap trap. Our podcast is here to bring you relevant, under
reported current events, as well as in-depth discussions from
a scientific, critical, skeptical, and humorous point of view.

Derek and Swoopy are your hosts, filling your pod and your
brain with skeptical insight and conversation, sometimes
heated, on a plethora of topics that are ripe for critical
examination. Bringing truth to podcasting, and all who choose
to listen.

In our travels we will tackle the beasts of pseudoscience; the
paranormal, supernatural, ufo / alien encounters, mis-
understood history, and overwrought legends – urban or
otherwise.

Our interview shows feature notable skeptics including;
leading astronomy speakers, scientists, philosophy experts,
and other scientific, secular, humanist, and skeptical book
authors, and critical thinkers.
```

Figure 1-7. The description of a podcast.

TIP

iTunes' Podcast Directory is not the only one out there.
Several director web sites were set up long before Apple
leapt into the podcasting fray, and some of them contain
more podcasts than iTunes.

Subscribing to Podcasts

So you've found a few podcasts you want to listen to: a great
science show, a podcast about music, one about snail racing,
or maybe the show where someone reads a page of the dic-
tionary every day. It's time to subscribe to the podcasts you
want to hear. (Or, simply download an episode to try out a

podcast; if you want to just download one show, see the section "Downloading Individual Shows," later in this chapter.)

iTunes is nothing more than a conduit; when you subscribe to a podcast with iTunes, you simply add a link in your iTunes library to the program's web site (this all happens under the hood). This web site provides the podcast and each new episode, even though iTunes grabs the podcast and gets it to your computer.

You've seen the Subscribe button in several of the figures earlier in this chapter. When you click one of these buttons, iTunes displays an alert, as in Figure 1-8, asking if you are sure that you want to subscribe. Click Subscribe to confirm your subscription.

Figure 1-8. iTunes checks to make sure you haven't clicked a button accidentally. If you don't want iTunes to harass you when you click Subscribe, check the box in the subscription alert telling the program not to ask again.

After you click the Subscribe button, iTunes takes you to the Podcasts list, adds the show, and starts downloading the latest episode.

Since your subscription only downloads the latest show, you may want to check out other shows available for the same podcast. Click the triangle icon next to the podcast's name to see all the shows you can grab from the program's archives (see Figure 1-9).

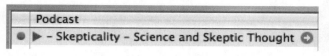

Figure 1-9. The blue dot icon to the left of the podcast's name indicates that it contains an unplayed show.

Downloading Individual Shows

While you can subscribe to a podcast and receive its new shows automatically, you may like it so much that you want to download some of its previous shows. The number of previous shows available depends on each podcast and how many shows it provides. (In some cases, you'll find only a few shows listed in iTunes, but the podcast's web site will have additional archives where you can download other shows.) If you want to find out more about a previous show, you can click the **i** icon to view a description. To download a show from this list, click the Get button.

You'll also see a Get Episode button on the iTunes Music Store page for each podcast. This button appears to the right of each available episode. Click this button to download an episode without subscribing. The podcast's name will be added to your Podcasts list, and a Subscribe button will appear next to this name. If you like the show you downloaded, you can click the Subscribe button to activate a subscription.

Updating Subscriptions

A podcast subscription is like a magazine subscription: just as you automatically receive each new issue of your favorite magazine in your mailbox, iTunes makes sure that you get new episodes of the podcasts you subscribe to as they become available. You can adjust the frequency of iTunes' updates, or you can update your subscriptions manually (either for all your podcasts at once or for an individual podcast) if you really need to get your fix.

If you click the Settings button at the bottom of the Podcasts window, iTunes opens its Podcasts preferences. (You can also view these preferences by selecting iTunes → Preferences on a Mac, or Edit → Preferences on Windows, then clicking the Podcasts icon.) Figure 1-10 shows these preferences.

Figure 1-10. The Podcasts preferences, where you choose how iTunes manages your podcast subscriptions.

While you can't choose subscription settings for individual podcasts, the Podcasts preferences let you choose how often you want iTunes to check for new shows, what to do when it finds new shows, and how many to keep.

Checking for new episodes

The first pop-up menu, "Check for new episodes," lets you choose from "Every hour," "Every day," "Every week," or Manually. Choose "Every hour" if you want to have new podcasts throughout the day; if you have lots of subscriptions, you'll find new episodes at different times, as each podcast updates its programs at a different frequency. "Every day" is probably a good choice for most people, which is why it is selected by default. You can also choose Manually if you only want to check on your own. (See "Manually Subscribing to Podcasts," the next section, for instructions on doing this.)

Downloading new episodes

By default, iTunes only downloads the most recent episode for each of your podcast subscriptions. If you change the check frequency to, say, "Every week," you might want to change this setting to Download All. Of course, if you add a podcast with a large archive to your subscription list and select Download All, you may end up with many gigabytes of podcasts. This would, however, ensure that you get the entire series of snail-racing podcasts, so you may want to choose this if you find certain podcasts that you really like.

Keeping episodes

This menu lets you choose how many episodes to keep. By default, iTunes keeps all unplayed episodes, which means that after you have listened to an episode, iTunes deletes it. But note that iTunes marks podcasts as played even if you've just started listening to them. You may, therefore, want to change this to All Episodes, then manually delete the ones you've finished listening to and don't want to keep. (See "Organizing and Saving Podcasts," later in this chapter, for more on keeping podcasts in your iTunes library.) Other choices here include Most Recent Episode, Last 2 Episodes, and so on, up to the Last 10 Episodes.

Manually Subscribing to Podcasts

We mentioned earlier that iTunes is the "one-stop shop for podcasts," but we also pointed out that there are podcasts that aren't listed in the iTunes Podcast Directory—lots of them, in fact. You're not limited to subscribing to podcasts that are in the iTunes Podcast Directory; you can manually subscribe to any podcast, as long as you know its URL.

A *URL*, or *uniform resource locator*, is a web page address; most podcast URLs end in *.xml* or begin with *feed:*. If you find a podcast you want to add to your subscription list, look for a link on the podcast's web site. From your web browser, copy the link (you'll usually right-click in Windows, or Control-click on a Mac, to display a contextual menu, and then select a command such as Copy Link or Copy Address).

Next, switch to iTunes and select Advanced → Subscribe to Podcast. Paste the URL into the dialog (shown in Figure 1-11), and then click OK.

Figure 1-11. Manually adding a podcast to a subscription list.

iTunes then checks the web site, downloads the latest episode, and lists any other episodes that show up on the site. From here on, just follow the previous instructions for managing your subscriptions.

Unsubscribing from Podcasts

Egad! You've subscribed to 327 podcasts, and you suddenly realize that you don't have time to listen to 86 hours of talk every day. (This makes sense, of course; there aren't that many hours in Earth days, though if you're on Venus, you'll have 243 hours in every day to work, sleep, and listen to your favorite podcasts.) You'll quickly have to prune down your list to fit your amount of listening time.

There are two ways to expunge podcasts from your subscription list. The first, more radical method, is to simply select a podcast (the program name, not an individual episode) and press the Delete key. If you've got any episodes of the show on your computer, iTunes will ask you if you want to move them to the trash or keep them in your iTunes Music Folder. If you choose to keep the files, they are curiously not available from your iTunes library, but you can navigate to your iTunes Music folder and its Podcasts subfolder if you want to find them. But if you move them to the trash, they get nuked. You'll probably choose this latter option most of the time because if you don't like a podcast, you won't want to keep its files on your computer.

The second way to unsubscribe from a podcast is to select it in the Podcasts list, then click the Unsubscribe button at the bottom of the iTunes window. This keeps the podcast in your Podcasts list and keeps any shows you've already downloaded. As you can see in Figure 1-12, it also adds a Subscribe button next to the name of the podcast; if you ever want to resubscribe, just click this button.

Podcast
● ▶ – Skepticality – Science and Skeptic Thought ◯
● ▶ iTunes New Music Tuesday   (SUBSCRIBE) ◯
● ▶ TheMusicNeverStopped.net ◯

Figure 1-12. After you unsubscribe from a podcast, you can click the Subscribe button to start up your subscription again.

iTunes keeps shows from all podcasts in the Podcasts list, whether you are subscribed or not, according to the settings you have chosen. See "Updating Subscriptions," earlier in this chapter, for information about these settings.

iTunes is also smart enough to warn you when you haven't listened to a specific podcast for a while. If so, it displays an exclamation point in a gray circle next to the podcast's name (where the blue dot would be). Click this icon to display an alert, such as in Figure 1-13.

iTunes has stopped updating this podcast because you have not listened to any episodes recently. Would you like to resume updating this podcast?

Cancel   Yes

Figure 1-13. iTunes stops updating podcasts you haven't listened to recently. You can restart updating by clicking Yes.

Click Yes to resume updating the podcast. Or, if you don't want iTunes to update it, click Cancel.

Deleting Individual Shows

So you're running out of disk space on your computer, but you still want to save a lot of your favorite podcasts to listen to again. You may want to cull your podcast list to save space, without deleting entire podcasts or all the shows you've listened to. After all, when snail-racing season is over, how else can you relive the thrills and excitement of those sprints to the finish line?

You can delete individual shows by selecting a show, then pressing the trusty Delete key. iTunes asks if you want to keep the shows or not; generally, if you are deleting shows to save space, you'll choose to move them to the trash.

However, when you delete a show, it no longer appears in the list of episodes for the podcast. If you ever want to get it back, you can delete your subscription and resubscribe, but if the show is no longer in the podcast's feed, you won't be able to download it again.

You can also delete dimmed podcasts from this list, if their descriptions don't interest you and you think you'll never want to listen to them, and if you find they clutter your Podcasts list. Select one and press Delete; there's no alert for this.

Organizing and Saving Podcasts

For most podcast fans, Apple's organization of podcasts is sufficient. The Podcasts list, easily accessible from the iTunes Source list, gives you one-click access to your entire library of podcasts. Because podcasts are organized by name, with individual episodes hidden unless you click the disclosure triangle to see them, you could have hundreds of shows, yet still see only a few dozen podcast names in the list.

So, you may never need to change anything about this organization. If you have a recent enough iPod (4th generation or later, or any iPod Mini), you'll find all your podcasts in the special Podcasts menu; if you have an older iPod, podcasts show up in a Podcasts playlist. (See "Listening to Podcasts on Your iPod," later in this chapter, for more on this.) If, however, you have an older iPod and want to organize your podcasts in a different way, or if you're a control freak, you may want to make playlists for some or all of your podcasts.

For starters, iTunes handles podcasts differently than music. Podcasts don't appear in your iTunes music library, and all your podcasts are organized, on your computer, in a Podcasts folder, which contains subfolders named for each podcast. Because the files are not in your iTunes library, you cannot search for them in your iTunes library, nor can you create smart playlists of your podcasts *unless* you add your podcasts to your iTunes library.

You can create playlists of your podcasts, which is especially useful if you listen to your iPod in your car and want to organize your day's listening without having to go back to the iPod and find the next one you want to listen to with one hand while steering with the other. Just select File → New Playlist, or click the + button below the Source list, and a new, empty playlist is added to this list. Name the playlist (anything is better than Untitled Playlist); the playlist is highlighted so you can just type in a name for it. (To change the name, select the playlist and press Enter, then type a new name.)

Next, go to the Podcasts list and drag the episodes or podcasts you want to the playlist. You can add individual episodes by clicking the disclosure triangle next to a podcast and dragging the episodes you want, or you can add all the episodes of a podcast by dragging its name to the playlist. You can then reorder the podcasts by dragging them up or down in the list (the lefthand column in the playlist must be highlighted for you to move the podcasts; if this is not the case, click it to select it and sort manually).

After you sync your iPod (see the upcoming section, "Syncing Podcasts to Your iPod"), the playlist shows up in your Playlists menu.

Saving Podcasts

If you want to save some of your podcasts, the best way to do so is to copy them from the Podcasts folder, which is in your iTunes Music folder, and place them in another location. Because iTunes manages your podcast subscriptions and all the shows you download from the Podcasts list, you may inadvertently delete podcasts that you intended to save. And many podcasts, especially interviews or music podcasts—not to mention the 2005 snail-racing season play-by-play podcasts—are worth keeping for another listen.

The easiest way to find where a podcast's file is hiding is to right-click (Windows) or Control-click (Mac) the podcast you want to keep, then select Show Song File from the contextual menu. This opens a new window showing the file in the Finder (Mac) or in Windows Explorer. Copy the file to another folder so you can keep it. You can then archive the podcasts on CD or DVD, copy them to other computers, or send them by email to everyone in your address book. (Um, no. Don't do that.)

Syncing Podcasts to Your iPod

The word "podcast" includes "pod." And the word "pod" comes from...you guessed it: iPod. (But you've already seen that in the Introduction.) The ideal adjunct to iTunes, so you can listen to your podcasts on the go, is an iPod. (However, you don't need an iPod to listen to podcasts; you can use other music players to listen to them, or even use iTunes to listen on your computer.)

iTunes offers many options for synchronizing music from your iTunes music library to your iPod and also has a set of options for syncing podcasts. With your iPod connected to your computer, open the iTunes preferences (iTunes → Preferences on a Mac; Edit → Preferences on Windows), then click the iPod icon. Click the Podcasts tab to see a window similar to Figure 1-14.

Here's how you can sync your podcasts to your iPod:

To automatically update all your podcasts
    Check the first option, "Automatically update all Podcasts." When you connect your iPod to your computer, iTunes syncs all the podcasts you have, but only syncs episodes according to the choice you make from the Update menu.

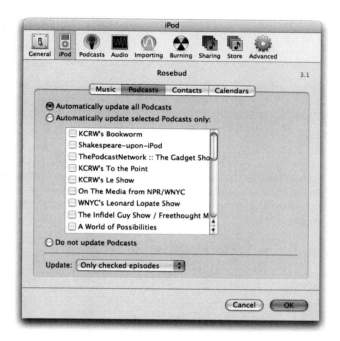

Figure 1-14. Podcast syncing options for an iPod.

To choose which episodes to update

From the Update menu, select All Episodes, "Only checked episodes," "Only most recent episodes," or "Only unplayed episodes." We mentioned earlier how you can uncheck certain podcast episodes; if you select "Only checked episodes," these unchecked episodes won't get synced. And if you select "Only unplayed episodes," only those podcasts you have not started listening to will get synced. (You may recall, from earlier in this chapter, that iTunes and the iPod consider an episode "played" even if you have listened to only a few seconds of it.)

To only update selected podcasts
> Check the "Automatically update selected Podcasts only" option. Then check the podcasts you want updated automatically. Others won't get updated.

To not update podcasts
> Check the "Do not update Podcasts" option. You can then manually drag podcasts onto your iPod to add them to its library, if you manually sync your music, or you can simply use iTunes to listen to your podcasts.

The way you choose to sync podcasts will depend on how many podcasts you listen to, but also on how much space you have on your iPod. If you have a small-capacity iPod, such as an iPod Mini, you'll need to share its space with your music library.

Syncing Podcasts to Other Music Players

Some of you may not have iPods. That's all right. Other digital music players can play digital music—and they can even play podcasts, as long as the podcasts are in a format the devices can read.

To sync your podcasts to a non-iPod music player, you need to know where iTunes stores your podcasts. You may need to point your music player's software to this folder, or you may need to copy podcasts manually from this folder to your player.

- On Windows, iTunes stores podcasts in *My Documents\ MyMusic\iTunes\iTunes Music\Podcasts*.
- On a Mac, iTunes stores podcasts in *Music/iTunes/iTunes Music/Podcasts*.

In both cases, podcasts could be in a different folder if you selected a different location for your iTunes Music Folder. (You can set this in iTunes' Advanced Preferences.)

The Podcasts folder contains subfolders, one for each podcast and named accordingly.

Listening to Podcasts

It's been a long haul to get to the listening section. Actually, subscribing to a podcast is so easy that you may have skipped most of this chapter to get to this section so you can listen to your first podcast. Here's how you do it.

Listening to Podcasts with iTunes

Do you want to listen to your podcasts on your computer with iTunes? You'll naturally need speakers—either built-in speakers or external speakers connected to your computer. Once you've got that set up, here's what you need to do:

To listen to a podcast
Select the podcast in the Podcasts list, then click the Play button at the top left of the iTunes window.

To stop listening to a podcast
Click the Pause button.

That's pretty easy, isn't it? Almost too easy. In fact, since it's that simple, it makes us feel somewhat useless. After all, do you really need us to tell you how to click one button to start or stop a podcast?

We like to feel useful, though, and in order to enhance our self-image, we're going to give you some tips about listening to podcasts with iTunes. These are actually great tips, so make sure you read them. Please?

You can start listening to a podcast by double-clicking it; you can also start listening by selecting the podcast, then pressing the spacebar. Press the spacebar again to pause the podcast.

You can either click the disclosure triangle next to the name of a podcast, and then select the episode you want to listen to; or you can select the podcast's name and click Play, press the spacebar, or double-click. If you do this with the podcast's name selected, you'll start listening to the latest episode of the podcast. However, when you get to the end of the episode, iTunes won't start playing the next episode in the list. You'll have to start that up manually.

When you switch to another podcast, or to some music, you can come back to the podcast and pick up *from the place you left off*. This is very cool: podcast files are said to be *bookmarkable* because they keep your place. Not only can you start listening to a podcast with iTunes but if you sync it to your iPod, the iPod records the bookmark so you can continue listening from where you left off. The same goes for listening on the iPod: if you stop and then sync your iPod to your computer, iTunes records the bookmark as well, so you can continue listening on your computer.

If you listen to *enhanced* podcasts with iTunes, you get some extra special features. Enhanced podcasts are special types of podcasts that contain *chapters*—or markers that indicate when segments start—and they can also contain graphics for each chapter. Not many podcasts use chapters yet, partly because these podcasts must be AAC files, which limits their listenability on non-iPod devices. But you can subscribe to Apple's iTunes New Music Tuesday podcast to see how this works. When you start listening to an enhanced podcast, you'll see a little pop-up menu at the top of the iTunes window, as in Figure 1-15.

Figure 1-15. The Chapter pop-up menu, which appears only when you listen to an enhanced podcast.

Click this button to display the chapter menu. Figure 1-16 shows this menu and the different chapter markers for one of the iTunes New Music Tuesday podcasts.

Figure 1-16. Skip through an enhanced podcast by selecting a chapter from this menu.

There's even more integration with this particular podcast, as well as with Adam Curry's *PodFinder* podcast. Click the album art button (see Figure 1-17) to display the graphic linked to the chapter.

Clicking the link at the bottom of the graphic, such as in Figure 1-17, takes you somewhere else. In this example, it takes you to the album or song page in the iTunes Music Store, but enhanced podcasts can also have links to web pages.

Figure 1-17. With the chapter graphic visible, you can see the same graphics that appear in the chapter menu. In some podcasts, a clickable link at the bottom of the graphic takes you elsewhere, such as to a page in the iTunes Music Store.

Listening to Podcasts on Your iPod

When Apple released iTunes 4.9, they also released an updater for the iPod. For 4th generation iPods or later (those with click wheels), or for iPod Minis, this updater adds a Podcasts menu item in the Music menu. Select Music, then select Podcasts to see a list of the podcasts you have on your iPod. The titles of the podcasts scroll and, when you select a podcast, you'll see a list of episodes (whose titles also scroll, so you can see the full titles).

You start listening to a podcast the same way you listen to music on your iPod. First, go to your podcast. If you have a recent iPod and have updated it since podcasts came to iTunes, you can go to Music → Podcasts, then select the name of a podcast, then an episode. If your iPod is a bit older, select Music → Playlists → Podcasts, then the name of a podcast, then an episode. Or, if you have made your own podcasts playlist, select that playlist, then select the episode you want to listen to.

Press Play to start listening. You can press the Select button to access different functions while listening to podcasts. If you press the Select button once, you can scrub (i.e., fast-forward) through the show. If you press it a second time, you'll see the show's notes.

You saw earlier how you can work with enhanced podcasts when listening with iTunes; the iPod lets you work with these podcasts as well, skipping through the chapters as you do with iTunes. You'll be able to see whether a podcast is enhanced right away—the time line features little vertical lines within it that indicate the location of chapter marks. Start listening to an enhanced podcast, then press the Select button once, and you'll see the name of the chapter in a small font just above the time line. Press the Next button to skip to the next chapter or the Previous button to go back. As you do, the iPod skips to the chapter mark, displays the name of the chapter, and starts playing from that point. A second press of the Select button displays program notes.

As we pointed out before, podcasts are bookmarkable. This is the case on the iPod, as with iTunes. So if you stop listening to a podcast on your iPod and want to listen to some music, don't worry. Just come back to the podcast later and start playing it: you'll pick up right where you left off.

Working with Other Podcast Software

In the middle of 2004, Adam Curry hacked together some computer code to create the first dedicated podcast client. Since he was an ex-video jockey for MTV, his hair looked great, but since he wasn't an experienced coder, the result wasn't pretty. It was, however, a proof-of-concept that podcasting could work. After the idea caught on and the Internet was abuzz with the new craze, some real programmers jumped in to help with the task, and the iPodder program was born.

Today, the ease of use and stunning good looks of iTunes make Apple's program the best choice for most listeners, and for that reason, we've looked closely at iTunes in this chapter. But there are many other programs that can handle podcasts, including iPodder, iPodderX, PodNova, and Doppler.

All podcast clients have the same basic features: integrated directories, scheduled downloading, and iTunes playlist synchronization. (They all let you add podcasts to your iTunes music library, so you can sync the shows to your iPod. Remember, "podcast" contains "pod"....) However, some users might find other features appealing and may want to explore alternatives. These programs have pros and cons, but we see four reasons why someone might want to use a different program to manage podcasts: to export their subscription list (or import other subscription lists), to use one-click subscription, to use an alternative media player (i.e., not iTunes), or to use a different portable music player (i.e., not an iPod).

One thing you can't do with iTunes is export your podcast subscription list. While many users won't need this function, exporting your subscription list comes in handy if you want to change your podcast client or switch computers. The other clients we've mentioned provide an export Outline Processor Markup Language (OPML) option, which means that if you want to try a different client or switch computers, you can take your subscription list with you in a handy file and import it into another program. This way, you don't have to resubscribe to all your favorite podcasts. You can even send this list to friends so they can check out your favorite podcasts without searching for them.

The PodNova client synchronizes your subscription list to the PodNova web site. This means you can access your podcasts from any web browser, even when you're at work or on the road. It also means that if a podcaster has a PodNova button on his web site, you can simply click the icon and the show will be added to your list. This one-click subscription feature is neat, quick, and practical, and other podcast programs are adopting it as well.

Many podcast programs integrate with media players other than iTunes. If you don't have an iPod, or prefer different music software, it might make sense to try another client. Most Windows clients work out of the box with Windows Media Player, or they manage podcasts and play them on their own.

Finally, if you don't have an iPod, and iTunes doesn't play nice with your MP3 player, then you might want to try out a different program that offers better syncing.

Here are the main podcasting programs, other than iTunes:

iPodder (Windows, Mac, Linux)
    http://ipodder.sourceforge.net/

iPodderX (Windows and Mac)
    http://ipodderx.com/

PodNova (Windows and Mac)
    http://www.podnova.com/

Doppler (Windows)
    http://www.dopplerradio.net/

These programs provide podcasting tools for Windows, Mac, and Linux, but podcasting clients have been written for other platforms as well. You can go to the Podcasting News web site and view the clients for many other platforms, such as the Pocket PC and Amiga: *http://www.podcastingnews.com/topics/Podcast_Software.html*.

CHAPTER 2

Starting Out in Podcasting

*Podcasting gives you the power to compete with
Howard Stern, from your basement.*

—Joe Lipscomb

An hour from now you can be a podcaster. It's far easier than
you think, and all you need are the microphone on your laptop and a connection to the Internet. Getting started early is
very important. Podcasting is all about making mistakes and
learning from them to create better podcasts. So, start right
now and make your first podcast.

TIP

This chapter was excerpted in part from *Podcasting Hacks*
(O'Reilly), which includes many tips and tricks for creating your own podcasts.

Make Your First Podcast

You can use the hardware you have right now, and some free
software on the Web, to make your first podcast.

If you don't have an internal microphone in your computer,
you will need to get a microphone. Microphone solutions are
available for all budgets. *Podcasting Hacks* covers the basics
of choosing and using microphones in detail.

Once you have the sound input device covered, the next step is to download Audacity (*http://audacity.sf.net/*), a free application that runs on Macintosh, Windows, and Linux. It can record sound from any source, including the internal microphone on your PC or Macintosh laptop.

With Audacity installed, press the big red Record button and explain what you have in mind for your podcast. The meter bars attached to the window will show you when you are talking too loudly (by hitting the far side of the meter near the 0 mark) or too softly (by registering only slightly as you talk). Click the Stop button to finish the recording. When you are finished, you will have something that looks like Figure 2-1.

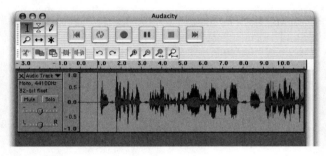

Figure 2-1. A recording in Audacity.

As the recording is made, your voice is shown as a waveform on the display. Each word you say appears as a little blip in the signal that goes above and below the center line. The louder the word, the taller the blip.

Figure 2-1 shows a short period of silence at the beginning of the recording. I didn't start speaking until one second after I pressed the Record button. You can remove that period of silence by using the Selection tool in the upper-lefthand corner of the window. Next, select the period of silence and click either the Delete key or the icon with the scissors to cut the signal. You can do the same at the end of the signal to remove any trailing silence.

Digital audio is exactly like digital photography or video, in that you can do as many takes as you like or do as much editing as you please. It's all just RAM or disk space, and you can delete what you don't use. So, relax and take as much time as you need to say what you want to say.

TIP

Being relaxed as you record your podcast is of primary importance in terms of getting a good sound. A number of handy tips for improving your vocal skills are outlined in *Podcasting Hacks*.

With your audio file edited, you need to save that Audacity file to disk, and then export the file as MP3. For voice-only podcasts, I recommend using a 32-bit compression rate for MP3. You set that in the Preferences dialog in the application's File Formats tab.

You will be prompted for some information about your MP3 file. This information is stored as ID3 tags that are embedded in your MP3. Getting the right content in those tags is important for making it easy for your listeners to find and listen to your podcast.

With the MP3 in hand, now you have to put it up on the Web and link it to a Really Simple Syndication (RSS) 2.0 feed. Several solutions are available for this, depending on what you have today.

You have no blog, domain, or ISP
If you don't have a blog, domain name, or ISP, the easiest way to put together a podcast is to use Ourmedia (*http://www.ourmedia.org/*). Other options include Liberated Syndication, Odeo, and AudioBlog.

You have a domain, but no blog

If you have your own domain and ISP, you have several options. You can set up a blog using Movable Type or WordPress.

You have a domain, but don't want to run a blog

You can podcast without a blog using Podcastamatic or Dircaster.

With most of these solutions (except for Liberated Syndication, Odeo, AudioBlog, and Ourmedia), you are going to have to find a place to host your MP3 file on the Web. MP3 files are a lot bigger than text files, so finding a place to put them can be difficult.

Once you have used one of these options to get your blog set up with an RSS 2.0 feed, and you've uploaded an MP3 file, you can create a new entry in the blog that points to your MP3 file. At that point, you are a podcaster! Point your podcatcher to your RSS 2.0 feed to make sure it downloads properly. Then go out into the podcastosphere and promote your podcast.

Where to Go from Here

The technical aspect of podcasting is only one part of the story. Once the microphone is switched on and the digital reels are virtually recording, what do you say? Content is king. Chapter 3 will give you some ideas about what to say and how to say it.

Still, the technical side of podcasting is a blast. You have a wide variety of amateur and professional microphones to choose from. Digitizing hardware is inexpensive and can greatly improve your sound. You can use cheap mixing boards and portable recorders to podcast anywhere, from your car to your local bar. The digital revolution has dropped the barrier to entry for communications, photography, and digital video. However, audio was not left behind. Chapter 4 covers some of your software options in detail.

Becoming a Critical Listener

Now that you are a podcaster, you will need to develop a new way of listening to podcasts. Instead of just lying back and enjoying, which you can still do, now you have to become a critical observer of podcasts. You are listening for several things:

Structure

What is the show's format? What recurring elements, called *format elements*, does the show use to keep you listening to this podcast and coming back for future podcasts? Is the interesting stuff in the beginning, at the end, or mixed throughout?

Style

How are they presenting themselves? Are they professional or aloof? Are they just goofing around? Is their style related to what they are talking about?

Technical elements

Are they using their blog in a unique or novel way? Have they put together something new with RSS? Do they offer a new way of contacting them with feedback? You should be on the lookout for all of these things when determining what to include in your show.

Content

What's holding your attention? This is particularly important because it's primarily what keeps people coming back to the show. When something moves you, listen to it over and over and figure out what is keeping you engaged.

You can learn from what does and doesn't work. When you hear something that works, you will want to take that idea and see if it can work on your show. And when something falls over, you will want to make sure you aren't making the same mistakes.

This pertains not only to podcasts but also to anything on the radio, on television, or in what you read. The structures remain the same throughout. The narrative arc that moves you in a 30-second commercial can also work in your podcast.

Think of yourself as a kid in Dad's workshop, taking apart a transistor radio to see how it works. You used to just listen to the radio, but now you want to see how it works and try to make it better. Podcasts are just like little machines that you can dig into and see how they work, and then apply those lessons to your own podcast.

CHAPTER 3

Formats for Your Podcast

The answer gets to the heart of what makes narrative work:
whenever there's a sequence of events—this happened, then
that happened, then this happened—we inevitably want to
find out what happened next. Also, and this is key, this banal
sequence has raised the question, namely, what's this guy
saying? And you'll probably stick around to find out.

—Ira Glass

This chapter covers the role of format in podcasts, starting
with an in-depth look at the why and how of formats. Then
some examples are presented that dig into the individual format types, and some case studies are offered along the way.
Several more examples can be found in *Podcasting Hacks*.

Adopt a Format for Your Podcast

You can apply the elements of a format to your podcast to
give listeners a reason to subscribe to your show.

To format or not to format? Many podcasters ask this question. Some believe formats smack of radio and are completely
inappropriate to the ad hoc podcast; others believe a format
can help get content to listeners in the best way possible.

Deciding if a format is right for your show starts with understanding the term.

In its broadest sense, the term *format* refers to a show's style. There are sports formats, talk formats, news formats, and others.

But, more specifically, the term *format* refers to how material within a show is *arranged*. In that sense, a format is an invisible framework on which your content rests. For example, you can format a sports show in different ways: you can feature a series of three quick interviews separated by music clips, or you can feature one long interview bookended by music clips. Both are sports shows, but they are formatted differently. How you arrange those blocks of sound is how you "format" your show.

Formatting starts with choosing an overarching theme for the show's content. Will it be a political show, a review show, a music show, etc.?

Once you have a theme, envision your ideal listener. Start with yourself: how would you like to hear the theme approached? With one guest, or with many guests? With lots of music and not much talking, or the other way around? If you're doing a music show, for example, you could play a lot of music and occasionally interview a musician, or you could host a talk show about music and feature an occasional performance by a guest.

Ultimately, some details of your format will change with each podcast, and others will stay the same. All that's important in this initial planning stage is that you decide what your listeners will enjoy hearing.

Decide on a Duration

How long should you make your podcast?

First, think about your listeners. What do you want to tell them in an episode of your show? How long will that topic hold your listeners' attention? Ask a few friends interested in your subject matter how long they might invest in listening to a show such as the one you're developing. Also, your listeners are likely to be doing something else while the show plays in their ears, such as working out or commuting. You can always match the length of the show to the length of that supposed activity.

Having some boundaries for the duration of the show is important for listeners because it lets them know what to expect. It's also important for you because it helps you to know how much content to put together.

Podcasts generally range from 15 minutes at the short end to 40 minutes at the long end. There are no hard and fast rules about time in podcasts; that's one of the great aspects of this medium. Typically, you want to start shorter and then go longer as you build your experience, or if the day's topic warrants it. One technical limit is 80 minutes—beyond 80 minutes and your listeners will no longer be able to burn a CD to listen to your show.

Another reason to decide on a duration for the show is to ensure you have enough time to produce it. An average podcast takes up to eight times the duration of the show to produce. A 15-minute podcast will take up to 2 hours to research, script, record, edit, and post. Shows that are mostly talk will take less time, and complex music or interview shows will take more time. You should know there are tradeoffs when it comes to production time—for example, you can save some time by not writing a script, but the unscripted recording might take longer to edit.

TIP

Here is a quick formula for determining duration: start with the number of free hours you are willing to commit to your show each week. Then divide that number by 6 and you will get a general duration for your show.

Block Out the Show

Radio shows map out their format based on a clock. They literally have a clock on a piece of paper, similar to Figure 3-1, which shows how long each segment of the show will be and what it will contain.

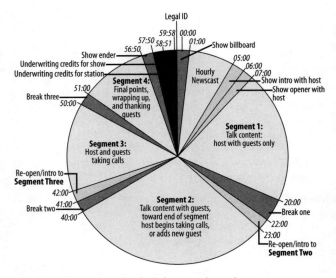

Figure 3-1. A simple radio clock for a one hour show.

A show is made up of blocks of sound known as *elements*. *Production elements* are standardized elements that are generally the same in each episode. These include such things as theme songs, introductions, and credits, and the way they're presented helps establish the show's identity. For example, using upbeat production elements creates the sound of a fun show.

Content elements, also known as *segments*, are the real substance of the program. These can be interviews, in-depth reports, or sets of music.

Between segments, radio relies on *transition elements*, which carry the listeners' ears from one piece to another. Usually these are bits of music, sound effects, or short bits of talk introducing the next radio segment.

But that's radio. Can these ideas apply in podcasting?

Elements offer predictability. They give a structure to the show that listeners can depend on, while allowing the host to be creative within that structure.

If you are already podcasting, you're probably using some basic elements. You likely have an introduction to start every show. And you probably have a farewell (a production element known as an *ender*) to wrap up your podcast.

But you can also use elements to set your podcast apart from others, by coming up with some regular segments that are unique to your show.

For example, David Letterman and other late-night television hosts have opening monologues. The audience tunes in, expecting the show to follow a certain format. But Letterman also has his "Top Ten List." When Dave moved from NBC to CBS, the contract specified that CBS would get the Top Ten List property. That's because a portion of the viewing audience shows up every night looking for the Top Ten List, even if they don't stick around for the rest of the show. The monologue is a standard content element in late-night talk shows, and the Top Ten List is a content element unique to Dave's show.

Format elements are the not-so-secret weapon that broadcasters (and narrowcasters) use to engage and hold an audience.

Create Great Content Segments

A great content segment does two things:

- It engages the listener by sounding interesting. This means if you're telling a story, you'll use an appropriate tone of voice (e.g., somber for a serious story), vary your pacing, and use dynamic language or good writing to keep listeners involved. If you're playing music, you'll put the songs in an appealing sequence, vary the rhythm or style a little or a lot, and create an evocative mood for the show. The way your content is presented *aurally* helps to engage your listeners.

- It informs, entertains, or inspires. What do you have to offer that's unique? Figure that out and build your show around it. Perhaps your news show reveals behind-the-scenes details your listeners don't have access to. Your talk show might feature someone with an unusual personal story, prompting your listeners to think about an issue in a different way or inspiring them to take action. If you're presenting an opinion, try to offer a fresh and well-reasoned perspective.

To develop great content, ask yourself if you'd listen to this segment, and be honest. Would it keep you coming back? Test it on a few friends or set up an email address and ask your listeners for feedback directly. Don't get stuck thinking you have to include things you don't care about. If you find yourself wanting to listen to a segment you've produced over and over, you know you've hit the mark.

Overall, great content *can* be listened to and enjoyed more than once. It's unique and universal at the same time. If it's engaging, your listeners will want to share it with others. And if you consistently create great content, your listeners will keep coming back, and their word of mouth will help build your audience.

Create Great Production Elements

At the very least, podcasts need an introduction and an ender. In the introduction, you introduce the hosts, give the name of the show, and offer a catchy phrase that describes what the show is about. Your ender should have whatever credits and sponsorship information you want to share with listeners. You should also thank the audience for listening, ask for feedback, and let them know they can find out more on your web site.

There are other production elements as well. A *billboard* or *rundown* is a short element that acts like an appealing contents listing for your show. Keep it conversational and describe what you are going to talk about in the order it will appear in the show. In this way, you encourage listeners to stay with you for the entire podcast.

Another production element is a *teaser*. This offers just a little information to get listeners excited about a segment that will appear later in the show. That works well on radio, but not as well in a podcast, where the person can just fast-forward to the segment. What *does* work is teasing your next show at the end of the current show.

Another production element is a *promotional spot* or *promo*. This is an element that isn't *within* your show, but instead, takes the form of a short (30-second or so) commercial for your podcast. It contains a clip from the show or a teaser that will get people excited. Ideally, these should be scripted and tightly edited. Promos come in handy when you want to trade spots with another podcast to attract more listeners or when you want to show up on Podcast Bunker (*http://podcastbunker.com/*).

A Few Lessons Learned

While researching this chapter, we had a chance to talk with a number of podcasters about what works and what doesn't. Here are some fundamental lessons they have learned:

Experiment

Try out new format elements continuously. Request feedback from your listeners and use this feedback to dictate how you do your show. Everyone we talked to said that with every podcast, they were improving. Strive to do the same.

Keep it light and have fun

Listeners can tell if you are bored, angry, or uninterested. Listening to your podcast is optional, and normal people don't stick around for something unpleasant. Follow Ben and Jerry's motto: "If it's not fun, why do it?" Listeners look at podcasters as friends, and who wants friends who are angry or unpleasant?

Respect the audience

If you think your time is valuable, you are not alone. Your listeners think their time is valuable, too. Listening to your podcast needs to be a rewarding experience for them. So, spend the time to design and write a podcast that provides a unique perspective and is worth the audience's time.

Strive for saucer eyes

Keeping your audience engaged is critical to keeping them coming back. Broadcasters call the effect of total engagement *saucer eyes* because of the look that people give you when they can't wait to hear what's next. If you hear a story that holds your interest, break it down and ask yourself what you liked about the content, and what about the way it was told kept you enthralled.

Keep it simple

Storytelling through audio works best when you have a linear narrative flow from the beginning to the end, and when you have only a few characters and key themes. Have a few larger points you want to get across in your podcast story, and keep returning to them and reinforcing them. A story should have a beginning, a middle, and an end.

Multiple hosts

Adam Curry makes the single-host format look drop-dead easy. The reality is quite different for the average human who doesn't have years of MTV experience. A multihost format is much easier to pull off because you can think about what to say next as your co-host is talking (of course, be sure to listen actively enough to your co-host so that you can respond appropriately). It's also easier on listeners to hear two people have a conversation than it is to listen to one person rant. So, grab a buddy and podcast together.

Remember, the most unique and important part of your show is *you*. Tell us what *you* care about, give us *your* thoughts in your own voice, and do the show *your* way. If you care about what you're saying, we will, too. Be yourself: that's what will keep it interesting for listeners.

The next section includes an example of a podcast format you can try on your own. More examples are covered in *Podcasting Hacks*.

Build a Great Story Show

Since January 2004, WGBH-FM has broadcast a weekly series of short radio pieces called Morning Stories, which are narratives from ordinary people about some significant moment in their lives.

The storytellers come from widely different backgrounds (they include a housecleaner from Brazil who is in the U.S. illegally, a postal worker, a tenured university professor, a retired high school teacher, a Cambodian immigrant, a blue-blooded Bostonian, a Jewish grandmother, victims of domestic violence, and seers of past lives). But their stories share a common goal: to give listeners a feeling for what it is like "to be the other guy."

In public radio, stories that make a strong personal connection with a listener are said to have *driveway potential* (i.e., they keep the listener glued to the radio, even after he has parked his car). At their best, these stories also make us want to retell them, in our own words, to someone else. Whatever their style or subject, stories this vivid and infectious tend to have some common elements:

- A dramatic story line that whets our appetite for what happens next
- Specific incidents, behaviors, and details that are easy to visualize (not, for example, "Full of self-reproach, I took it out on others," but "I came home and kicked the dog")
- A narrator with a genuine tone of voice, able to express or suggest the emotions the story inspires

Most of us are not professional raconteurs, able to tell a great personal anecdote to a stranger on demand; so, how do you get people with little or no radio experience to talk on the radio about something significant that happened to them in a way you just can't forget?

Basically, get them to feel that the experience they want to tell us is happening to them, right then and there. We are all born storytellers, hardwired to make sense of important experiences by shaping and passing them on in story form. The closer we get to the heat of the moment that our story is about, the more lively, compelling, and relevant the details of our account are likely to be.

At Morning Stories, we have learned that one of the most effective ways of getting vivid, first-person accounts is through an hour-long recorded "talk" with the producer. Starting with the topic or experience she wants to explore, the storyteller is encouraged to re-experience some of the moments and feelings that most moved her in the past and that move her now. In the course of the hour, we can usually get more than enough images and incidents from which a three- to six-minute story can be edited. On occasion, a fully formed story takes shape on its own.

Talk might be the wrong word to use because, in this process, the main job of the producer is to listen—actively, with all his senses and imagination—for what the teller says that strikes a genuine human response and for what the teller has not yet said that might make the story take full shape and come alive.

Experience helps, of course. But there are some skills you can learn that are specific to the art of interviewing for a story.

Listen for Specifics

Is the teller giving you images and details you can respond to or that you can flesh out with memories and experiences of your own?

One of the great powers of audio as a medium is that it involves us in finishing the picture with sensory memories and images from our own experiences. This is part of the process that lets us turn someone else's story into our own.

The villain the storyteller brings to life in her story becomes real to each of us in our own way. Based on our experiences, a villain might have a mustache, a dark felt hat, and a limp; yours might not. But both are real.

Echo the Teller's Emotions

The teller's tone of voice is a crucial part of her story. It reflects the emotions, spoken and unspoken, that she feels. The more you can encourage her to experience the emotions her story brings up, the more real and richer her tone of voice will be.

Listen to what she is feeling as she speaks, reinforce the feeling if it seems vague, and when in doubt, make sure you've got her right. "You must have been scared" or "What an embarrassing situation" or "That makes me feel sad" are examples of responses that let the teller know you're listening to what she is feeling and that you want to know more.

"Listen" to Your Own Reactions

Regardless of whether you actually talk about it aloud, be aware of what in your own life the teller's story is bringing spontaneously to mind. Can you see, hear, smell, or feel what the person is describing in your mind, out of pieces from your own memories and experiences? Are you interested, vitally, in what happens next because it has touched on something that matters deeply to you? Is what she is saying perhaps leading to someplace you have already been? Is she leaving a clue that might lead you both to where the story wants to go?

We have discovered that most of the stories we end up using on Morning Stories are not the stories the tellers start out thinking they are going to tell. Rather, they are the stories that come to life as part of the process of talking and being listened to.

In a sense, during the hour they spend together, the story-teller and producer are on a journey through only partially mapped territory, with the teller leading and the producer one step behind. The teller's job is to report what she sees and feels, and the producer's job is to follow behind, to keep the storyteller from wandering off track. Trusting the process will get you both where you need to go.

CHAPTER 4

Editing Your Podcast

I don't know that I'm doing it so much as a protest against radio as I am to develop the radio show I always wanted to hear.

—Brian Ibbot

Broadly speaking, audio production software falls into one of four categories: recording, editing, mixing, and plug-ins. Recording tools sample audio from one or more inputs and store it to disk. Editing programs allow you to rearrange or delete sections of previously recorded audio. Mixing tools take sounds collected from various sources and mix them together into a final produced piece. Plug-ins can be used throughout the process to add or remove distortion, noise, and echo, and to create sonic effects.

If all you plan to do is record a mono track of speech, make some small edits, and then encode it to MP3, you only need one program, Audacity. For more complex audio work, professionals use a combination of multiple programs, each of which specializes in one of these tasks.

In addition to these basic tools, other types of specialized applications can be useful to podcasters. Small audio routing applications, such as Soundflower, can make it easy to reroute audio within your system from applications that might not have elaborate sound input and output options.

Cart applications can be a handy way to organize the samples and sounds that you will use throughout your show. These applications preload the sounds for easy access and then present a grid of buttons; you press a button to play the corresponding sound.

An altogether new type of tool is the podcasting application. This tool performs a number of functions, from holding the show notes, to providing cart application functionality, to recording and playing back segments.

Choose the Right Audio Tools

A wide variety of free and commercial audio tools are available for podcasting. But that doesn't mean it's easy to find the right software. You can spend a lot of money on an application that is great for musicians but doesn't have the right functions for podcasters.

In this part, we cover both the free and commercial tools, explaining which ones are good for podcasting, and why.

Audio Editors

Audio-editing applications allow you to edit sound as though you're using a word processor. You can cut, copy, and paste; delete; and arrange sound in any way you choose. Most of these applications allow you to work with multiple tracks, which you can think of as sonic layers. You can use these tracks to work with each sound in isolation, and then mix down to a single mono or stereo signal at the end. In addition, many of these applications allow you to apply effects to the volume or the character of the sound.

Audacity

Audacity (*http://audacity.sf.net/*) is a free sound-editing program that runs on Windows, Mac, and Linux. You should download Audacity and install it *right now*. It's an excellent starter program and possibly the only editing program you will ever need. Figure 4-1 shows the Audacity main window.

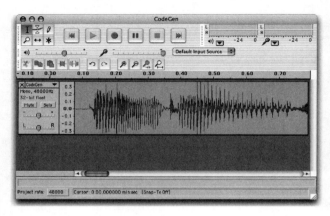

Figure 4-1. Audacity's main window.

On the recording side, Audacity supports mono and stereo recording from any sound input source on your machine. It even has meters in the document window to show when you are *clipping* (cutting off a portion of the signal). Unfortunately, plug-ins are not supported during recording. So, you will need to do all the filtering work after the recording.

Audacity has excellent editing capabilities. You can zoom and scroll on both the time axis and the amplitude axis. The keyboard controls for navigating around the document and then creating and extending your selections are easy to use. Cut, Copy, and Paste all work as if you were in a word processing document. However, copying and pasting between documents of different sampling rates can get tricky.

Audacity supports envelope editing. With this feature, you can manually boost or reduce the amplitude of parts of the recording using simple visual cues. This is very handy when you have sections of audio that drift from very soft to very loud.

When you download Audacity, you should also grab the LAME MP3 encoder and the Virtual Studio Technology (VST) plug-in support module. VST is a plug-in standard that is supported by a number of sound applications on both Windows and Macintosh. Audacity has a variety of filters already baked in, but the VST plug-in support expands your sonic toolkit tremendously.

Peak 4

BIAS's (*http://www.bias-inc.com/*) Peak 4 ($499) is one of the premier sound-editing tools on the market. It has a remarkably intuitive interface that allows you to find the samples you are looking for quickly in the alien world of waveform display.

Peak's sister application, Deck ($399), is a studio-quality recording application that supports mixer functionality with unlimited virtual software mixer channels. It's overkill for podcasters, but if you are creating a home studio for podcasting and music, you should check it out.

Audition

Adobe's Audition ($299) is on the Windows side of the recording and mixing cycle. Audition was originally Cool Edit, and then Cool Edit Pro, before it was acquired by Adobe. It's a professional's product—at an amateur's price—that's had years of honing.

With Audition you can multitrack record and edit. It offers a reasonable set of built-in effects, and you can extend the processing using VSTs. It has a high-quality tunable noise filter built-in. On the output side, a number of formats, including MP3, are supported.

On the downside, the interface is initially very complex and will be familiar only to those with experience in multitrack recording systems. On the upside, the documentation that comes with the product is an excellent introduction not only to the software but also to the science of audio recording.

Sound Forge

Sony's Sound Forge (*http://soundforge.com*) is a full-featured, multitrack recording and editing application for Windows.

A full complement of effects is built-in, but it also supports the VST plug-in standard so that you can add your own effects. An audio editor is built into the product, and the package includes noise reduction tools that will help you clean up subpar recordings.

If video is your thing, you can import video and sync up your audio editing with it. If creating music loops is more your style, this program integrates with Sony's ACID program. You create the loops here and then use ACID to choreograph them into a song.

At the time of this writing, Sound Forge sold for $299.95.

GarageBand 2

GarageBand 2, pictured in Figure 4-2, comes bundled with Apple's $79 iLife '05 suite. It's a steal alone at that price, and the inclusion of the other iLife applications makes it just that much sweeter. GarageBand is good for both recording and mixing sound.

Along the right side of the window are the tracks of the recording. Each row contains little blobs of audio. You can change their duration as well as their placement in time.

GarageBand doesn't support robust sound editing. So, you will still need an application such as Audacity or Peak to do your editing. However, GarageBand's ability to string together audio loops makes it ideal for making musical melodies for intros, outtros, bumpers, and stingers.

Logic Express 7

Apple's Logic Express 7 (*http://www.apple.com/logicexpress/*) is one step up from GarageBand 2. It's a sophisticated multitrack recording, editing, and production tool (see Figure 4-3).

Figure 4-2. GarageBand 2.

Figure 4-3. Logic Express on Macintosh.

The multitrack recording and editing capabilities are ideal for a home studio music recording setting. For podcasting, it will have way more functionality than you require, but at $299, it's hard to say no.

Sound-Recording Applications

A few applications specialize in the recording of sound. Some of these subspecialize in taking sound directly from applications in addition to traditional sources, such as microphones.

Audio Hijack Pro

Audio Hijack Pro (*http://rogueamoeba.com/audiohijackpro/*), $32, is a Macintosh podcaster's best friend. It's an application that can record audio from any sound input or any running application. This means you can integrate music or sound from iTunes directly into your recording. Figure 4-4 shows the Input tab of an Audio Hijack Pro session.

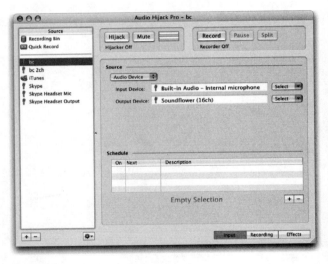

Figure 4-4. Audio Hijack Pro's Input tab.

On the lefthand side of the window is the list of sessions. Each session, as specified on the righthand side of the window, has an input and output source, details about where the output file is supposed to go (in the Effects tab shown in Figure 4-5), and the grid of effects.

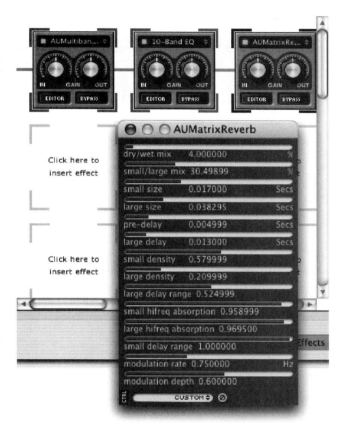

Figure 4-5. Audio Hijack Pro's Effects tab.

Audio Hijack supports a set of effects to alter the sound of your recording as you make it. These effects are strung together in an intuitive graphical format.

Through the Voice Over effect, you can bring audio from iTunes into your recording in real time. It's this ability to *hijack* sound from other applications—along with the application's incredible stability—that makes Audio Hijack Pro a very popular recording tool for Macintosh podcasters. Rogue Amoeba, the publisher of Audio Hijack Pro, has noticed this, and subsequent versions of the software are being built with podcasting in mind.

Total Sound Recorder

The inexpensive Total Sound Recorder (available at *http://highcriteria.com/*) is a favorite of Windows podcasters. The standard edition costs $11.95, and the professional version costs $39.95.

You can use the Total Sound Recorder Pro main window, shown in Figure 4-6, to record audio from all the standard input sources, as well as from applications.

Podcasting Applications

With the advent of podcasting have come specialized podcasting applications. These applications combine sound recording, rudimentary editing, and mixing, and they also automate the encoding and uploading process.

iPodcast Producer

iPodcast Producer is an all-in-one podcaster application for Windows. Figure 4-7 shows one recorded segment of audio.

Using this application, you can record your voice segments and drop in prerecorded effects and sounds that you can assign to hotkeys. A companion application enables you to edit the individual sound files to add some effects or to remove "ums" and "ahs" from your voice track.

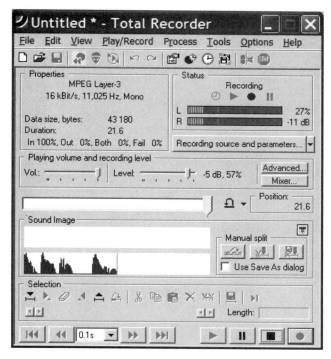

Figure 4-6. Total Sound Recorder Pro.

Once you are done developing the podcast, the application will mix it down to MP3 or WAV format for you. It will even go so far as to upload the mixed file using FTP and update your RSS 2.0 feed to add the podcast with the correct enclosure tags.

iPodcast Producer is available at *http://industrialaudiosoftware. com/* for $149.95.

Figure 4-7. iPodcast Producer.

MixCast Live

MixCast Live (*http://mixcastlive.com/*) is the first Macintosh podcasting application. It has a show notes area, a cart for samples, and a recorder built into the application. It was pre-released at the time of this writing. The purchase price is $59, but you can get a $20 discount if you want to take the leap early.

Fix Common Audio Problems

Here are some solutions to common audio problems. In audio you always have many ways to do a single thing, but you can use these as a starting point on the way to finding your own solution. Several more solutions are featured in *Podcasting Hacks*.

Reduce Wind Noise

Wind noise creates a loud rumbling that is below the 100 Hz level in recordings. You can use a low-pass filter or EQ to attenuate this effect. However, it's unlikely you will be able to remove the noise entirely. You should aim to reduce it to an unobtrusive background level and go from calling it *noise* to calling it *ambience* instead.

The ideal solution is to resample the sound with a windscreen and a filter on the microphone to take out wind noise.

Fix a Muffled Voice

If a person moves his mouth away from the central axis of the microphone, his voice will drop and become muffled. Try boosting the signal in ranges between 5 kHz to 7 kHz with an EQ to boost up the clarity of the voice. Then use a gain envelope to boost the signal to the level of the rest of the podcast.

Remove Hiss

Hiss is high-frequency periodic noise. The high frequency gives it the annoying high pitch. Thankfully, this also moves it away from the voice spectrum so that you can remove it without too much distortion to the original signal. Use an EQ to remove frequencies above 7 kHz. Use less EQ in the closer frequencies and be more aggressive in the higher frequencies.

Simulate a Phone

When you have one side of an interview recorded through the phone and the other side on a clean studio microphone, the result can be jarring to the ear. It helps to take a little of the quality out of the studio microphone recording by dropping off the low end below 60 Hz and reducing the high end above 3.5 kHz. Don't kill them entirely because then it will sound like you are on a phone; instead, reduce them a little to lessen the jarring difference in quality between the two sounds.

Another technique is to use a coffee mug on its side to trap part of your voice.

Simulate a Radio

Growing up in the mid-80s, one of my favorite songs was Wall of Voodoo's "Mexican Radio": "I wish I was in Tijuana, eating barbequed iguana." At several points in the song, the singer sounds like he is talking through an AM radio, severely boxed and distorted.

You can use digital effects to simulate this by doing a hard chop of the frequencies below 500 Hz and above 3 kHz with an EQ, and then over boosting the top end of the vocals around 3 kHz to add some distracting gritty clarity. Adding in another track of toned-down white noise will also add a little sonic grit.

If you have the time, another solution is to record through an actual radio. Put your recording on your iPod, and then use a Griffin iTrip to broadcast it on an FM band. Tune your radio close to that band. Analog tuners are better because you can get them close but not quite there. Move the iPod around until you get some randomly slight cut-outs. Then, record the signal coming out of the radio with your microphone. Using the internal microphone on your computer will add another level of grit and boxy compression.

Where to Go from Here

After you've chosen the format for your podcast, selected the tools to record and edit it, and done the real work of recording it, you'll need to package the file and post it on the Internet. The many options available for doing this are beyond the scope of this pocket guide. For detailed information about where to go next, we suggest you pick up a copy of *Podcasting Hacks* or check out the excellent free resources at Our Media (*http://www.ourmedia.org/*).

CHAPTER 5

30 Great Podcasts

In this chapter, we present our own highly subjective and by no means inclusive guide to some quirky, funny, informative, and, at times, even useful podcasts. Each of the podcasts in this chapter has one of the following ratings:

G: General Audience
> The podcast has great content for every listener.

PG: Parental Guidance
> The podcast might contain some language or content that requires guidance from an adult.

M: Mature Audiences
> The podcast contains language or content that might offend some listeners; recommended for adults only.

They run the gamut: from sports and tales of piloting and truck driving, to informative shows about digital photography, wine, beer, and having coffee with your friends and acquaintances. What they all share is a highly personal view of their subjects and a fascinating look into the lives of some extraordinary, interesting, and oddly knowledgeable people. You won't be disappointed. It's an amazing world out there, and these folks are illuminating it in charming and extremely thought-provoking ways. Happy hunting!

Friday Coffeeblogging

URL:	*http://www.candleboy.com/fridaycoffeeblogging/*
Frequency:	Weekly
Duration:	30 minutes
Rating:	PG

Friday Coffeeblogging is unique among podcasts and not only for its content: the show is recorded in a coffee shop in downtown Burlington, Vermont, and the atmosphere is recorded right along with the show. Captured in the background are the clinks of coffee cups, the murmur of the patrons, and the warbling of the café's background music. (And you can almost smell the coffee....)

The hosts of the show are Bill Simmon, N. Todd Pritsky, and Gregory Giordano, also known by their "web handles" Bill, Ntodd, and Flameape. In some respects, the show is like Seinfeld: it's about nothing. This podcast is simply an exploration of whatever happens on the day it's recorded. This lack of direction may turn you off, but the personality of the hosts is what really grabs you.

This podcast is a talkfest among three close friends with the occasional slice of music thrown in to spice things up, and, in most cases, the conversation slips into geek speak. Discussions range from meanderings and personal anecdotes, such as Flameape's claim that he is a personal disaster zone, to opinions of movies such as *Star Wars: Episode III—Revenge of the Sith* where the gang interviewed fans waiting in line for the premier at Burlington's Roxy Cinema.

The Bitterest Pill

URL:	*http://www.thebitterestpill.com/*
Frequency:	Almost biweekly
Duration:	20 minutes
Rating:	G

Being a semi-employed actor, comedian, writer, stay-at-home Dad, and a podcaster makes for some interesting tales. Having earned his acting stripes in the theater, commercials, and a handful of movies and TV shows, such as *The X Files* and *Chicago Hope*, Dan Klass tells stories of fatherhood, his auditions, and his recent venture writing a book about podcasting.

Klass is a veteran of using the Internet as an artistic medium, having recorded some of his stand-up routines for online distribution; with this experience, he is a consummate podcaster. His stories relate everyday family incidents, such as taking his sick five-year-old to the movies to escape the kitchen remodeling at home. Dan provides an entertaining impersonation of himself trying to control his two-year-old during the 80-minute showing of the movie *Madagascar*.

Dan's self-deprecating style and stories make for captivating monologues that keep you listening. He sometimes invites guests, like his son Hudson, or plays voicemail left on the Bitterest Pill listener line, and to round out the show, he plays some podsafe music selected from an online source.

Dr. Karl

URL: *http://www.abc.net.au/science/k2/*
Frequency: Weekly
Duration: 40 minutes
Rating: G

The world's most down-to-earth doctor, Karl Kruszelnicki, attracts 300,000 listeners a week on Australian Broadcasting Corporation's youth radio station Triple J. He hosts a question-and-answer show dedicated to science, where he answers incredibly vital questions such as, "Why is my nasal hair square?" and "How can I explode my own urine?" and "Why can I see through glass?" For years, Aussies phoned into Triple J asking thousands of scientific questions, and Dr. Karl provided the answers each Thursday.

Some questions on the show rate as bizarre, but many of them are the kind of things you always wanted to know but never really thought about. In some instances he regales listeners with stories from his multifaceted career, such as one where he proved the theory of mind over matter when he provided a placebo to a patient with kidney stones. The patient's body was so convinced that it had received a drug that when Kruszelnicki provided the real dose, the patient reacted like he had overdosed.

If you thirst for obscure scientific knowledge, the *Dr. Karl* podcast is a must listen. Check it out every week because no two shows are ever the same and we guarantee you'll learn something new each time.

Catholic Insider

URL:	*http://www.catholicinsider.com/*
Frequency:	Two or three per week
Duration:	15 minutes
Rating:	G

When Pope John Paul II died in April 2005, Father Roderick Vonhögen, a 36-year-old Catholic priest of the Archdiocese of Utrecht in The Netherlands, traveled to Rome and provided podcasts of the event as thousands of people converged on the Vatican to take part in the funeral.

In less than a week, over 10,000 people downloaded Vonhögen's subdued "Last Respects to John Paul II" podcast. Listeners could accompany him as he rose early and then rode on a World War II vintage bicycle to Saint Peter's Basilica to stand in line to pay his respects. It was a tour de force of audio treats, with the combination of the sounds of birds in the Italian night, mad Roman drivers, and whispers from the cavernous hall as people paid their respects.

Father Roderick's show is far from the smite-thee-down attitude one might expect. He's a present-day priest to the max, from the mock iPod silhouette splash page on his web site to his funky intro music and his jocular interview with Arnold Schwarzenegger, actor and governor of California. He shows the world that even a man of the cloth can have fun online.

Winecast

URL: *http://winecast.net/*

Frequency: Weekly

Duration: 15 minutes

Rating: G

If there's one thing we know about, it's wine. Kirk lives in France, the world's greatest wine-producing country, and Richard lives in Australia, which is fast becoming its biggest rival. So, it is with great drinking…um, listening pleasure that we tune in to *Winecast*. It's best to listen to *Winecast* with your favorite bottle of Sauternes, Shiraz, or Chardonnay; if not, you'll probably be cracking open a bottle soon after the show.

Tim Elliott from Minnesota, inspired by the *Evil Genius Chronicles* and *Reel Reviews*, started his podcast with the hope that he'd help wine drinkers climb out of the rut of buying the same type of wine time and again. After all, there are thousands of wines from dozens of countries—not only France and Australia—but understanding and choosing different wines takes a lot of experience.

If there's one thing Tim Elliott has, it's experience with wine. Focusing on a different type of wine for each show, he tastes the selection before he starts the show, or pops the bottle and slurps live, providing the listener with all the details without slurring a word. He even gives an overview of the region that produced the wine he tastes, along with some local wine history. From German Beerenauslese to California Zinfandels, he varies his choice each week.

Soundseeing Tours

URL:	*http://soundseeingtours.podshow.com/*
Frequency:	Contributed by a number of authors, so it varies
Duration:	Contributed by a number of authors, so it varies
Rating:	PG

Fancy a waltz through Disneyland, a trip to the Pacifico Market in Madrid, or a walk into Pacal's tomb in Mexico? Well, you can visit these places from the safety of your own iPod. Soundseeing is a term used to describe podcasts that take you to all parts of the world by recording the audio. In a collaborative effort prompted by Adam Curry, podcasters worldwide record a tour or journey they make to mundane or exotic locations. Whether it's the puffs and pants from inside a Canadian hockey player's helmet or hearing cars zip past a couple in Sri Lanka, there are plenty to choose from.

Each week a new batch of soundseeing podcasts are added to the collection. Anyone can contribute, which means that it is becoming an incredibly diverse collection of shows. Some are meant for the armchair traveler, and others are created to take with you as you explore a new country, town, or place.

Soundseeing Tours is a smorgasbord of global sounds, and each one can summon other senses—before you know it you can smell market fruits, feel the Sri Lankan heat, and see the British countryside.

Small World Podcast

URL:	*http://www.smallworldpodcast.com/*
Frequency:	Almost daily
Duration:	10 to 30 minutes
Rating:	M

The U.S. comedian Steven Wright once said with deadpan humor, "It's a small world but I wouldn't want to paint it." Bazooka Joe on the other hand wants to interview most of it. Bazooka, a DJ from Free Radio San Diego, the city's longest running and most notorious unlicensed radio station, spends several hours a week interviewing people for his daily show. His subjects range from a computer hacker involved with the Russian Mafia to a ballerina, and from a former male prostitute to a comic book artist.

Joe said that podcasting gives him an opportunity to look into other people's lives or cultures, but he doesn't just look, he delves. This habit might stem from his disc jockey background. Having dabbled in playing *agro* and *assault* music and working for a pirate radio station, he's always on the lookout for airing content from the fringe. So, interviewing people outside of standard typecasts is a natural progression, and he excels in finding fascinating stories.

Not for the xenophobic or faint of heart, the show is a fascinating exploration of today's cultures from different parts of the globe.

History According to Bob

URL:	http://www.summahistorica.com/
Frequency:	Daily except Sunday
Duration:	10 minutes
Rating:	G

Flaming Pigs is not the name of a punk rock band, but it is a history lesson by Professor Bob Packett, a teacher at Maple Woods Community College. Apparently, when elephants were commonly used in ancient battles, it became an element of war to smear flammable liquid pitch on pigs, light them on fire, and then release them into a battle to upset the war-trained elephants.

Bob describes himself as a historical generalist, and in his podcast he discusses many of the nooks of history that are not traditionally explored in classic lessons. These include topics such as Incan invincibility, the justice of pirates, and U-boat etiquette. It's these little-known snippets of history that make Bob's short lessons so fascinating.

Not satisfied with these historical oddities, Bob also provides a range of significant historical events, such as the fall of the Roman Empire or the Cuban Missile Crisis.

Bob's character is ever present through the show. It's apparent that he would talk for hours if he thought he could. When his wife suggested he keep the show to four minutes or less, he protested that he could hardly say hello in that amount of time. Fortunately for us, Bob talks for as long as he needs in the six shows a week. How else would we find out about flaming pigs?

Art Mobs

URL:	*http://mod.blogs.com/art_mobs/*
Frequency:	Varies
Duration:	Varies
Rating:	PG

A front-page article for the *New York Times* called *Art Mobs* irreverent. It's an accurate description; some of the shows are just shy of mocking the Museum of Modern Art (MoMA) and the works of art it contains. Developed by a class from Marymount Manhattan College, *Art Mobs* is a collection of podcasts that explore some of what you'll find captured in Manhattan's MoMA.

Several of the podcasts could be described as flirting commentary. Professor John Benton and two students, Malena-Amaranta Negrao and Cheryl Stoever, discuss the possible sexual undertones of a Cindy Sherman self-portrait and a Jackson Pollock painting.

Seeing the recordings as a form of respect for the museum, rather than derogatory, Gilbert encourages others to record audio to add to the collection. He hopes it will grow to encompass a large selection of the art in the museum, enough for a listener to take as a soundseeing podcast in replacement of the museum's rentable audio guides (visit *http://odeo.com/audio/81964/view*).

Good Beer Show

URL:	*http://goodbeershow.com/*
Frequency:	Weekly
Duration:	30 to 50 minutes
Rating:	PG

If your local pub stocked 300 different international beers at any one time, wouldn't you be compelled to try every beer in the world? Jeffrey T. Meyer, the host of the *Good Beer Show*, seems to have exactly that goal, and he's well on his way, having finished his 800th different beer for his 20th show.

The show is recorded live at the Heorot pub, which is named after the stronghold of King Hrothgar in the ancient epic poem *Beowulf*. Meyer wouldn't sound out of place in the mythical verse, with a gruff voice and an attitude about beer that would make a warrior proud. In fact he sums it up when he says, "I don't drink big brother beer," in reference to beers made by major brewing companies. He's more interested in fine independent brews.

Interviewing purveyors of other local pubs and boutique breweries, he and his rowdy posse focus on every type of beer imaginable. Each show focuses on a few featured beers. Meyer provides background stories, taste descriptions, and commentary while the crowd bubbles in the background. He knows his beers—he should, he's drunk enough of them. Beers range from Triple Karmeliet (Rich's favorite, if you ever need to know) to Samuel Adams Triple Bock—which he rates as the worst beer he's tried.

Skinny on Sports

URL:	http://skinnyonsports.podshow.com/
Frequency:	Weekly
Duration:	10 minutes
Rating:	G

In 2004, the world watched as the Indiana Pacers' and Detroit Pistons' basket-brawl spilled into the stands when a fan threw a beer cup at a player's head. Andy and Matt Skinn, brothers from Calgary, launched the podcast with an analysis of the NBA's reaction and their own opinions in a quick two-minute review. Ever since then, the brothers have provided a 10-minute sports news podcast each week at an unrelenting pace.

The show certainly is skinny, and justifies their claim of "the fastest 10 minutes on your iPod." Each show is split into approximately five equal sections: four quarters and over-time, just like a basketball game without the brawls. It's a to-the-point show with each quarter focusing on a different sporting event.

When Podcast Alley, an online podcast directory, inter-viewed the duo, Andy made the suggestion that *Skinny on Sports* is great even for those who are not into physical activ-ity. Being only 10 minutes in length, it's a quick way for any-one to bulk up on sporting knowledge. It could be used to impress work colleagues at the water cooler. Perhaps we should rename the show, *The 10-Minute Manager to Become a Sport Jock*.

Endurance Radio

URL:	*http://www.enduranceradio.com/*
Frequency:	Each Monday, Wednesday, and Friday
Duration:	15 minutes
Rating:	G

Tim Bourquin hosts three shows a week, where he speaks with athletes, coaches, and race directors who are involved with triathlons, adventure races, marathons, cycling, and swimming. Not only does he interview them about their inspirational stories, he also quizzes them about training methods, racing, dieting, and resting.

The show provides a collection of interviews covering a bunch of sporting angles: how Kate Major won the Lake Placid Ironman in 2004 after only four years training for the sport; how Pam Reed prepared for the heat in a desert ultra-marathon; and what Dave Scott, six-time Ironman World Champion, thinks about sports nutrition.

If you're concerned that your beloved iPod won't survive a triathlon (that's our excuse and we're sticking to it), take heart knowing that Bourquin's player has survived three mountain bike accidents. So with no other excuses left, download a show or two and head out the door for a walk, run, or quick triathlon.

Sports Bloggers Live

URL:	*http://www.sportsbloggerslive.com/*
Frequency:	Biweekly, Tuesday at 7 p.m. ET, and an extra midweek
Duration:	30 to 60 minutes
Rating:	G

When Jamie Mottram, a self-proclaimed citizen journalist, had a minute with Mike Tyson, he asked the boxer, "Have you ever thought of doing a reality TV show?" Tyson responded, "Never in my life because when I see reality shows, it's all about disrespecting and humiliation. I'm not interested. I've done enough of that in my life on my own and I wouldn't allow someone else to do that to me."

In his weblog, a supplement to the show, Mottram comments about the Tyson press conference, "A boxing press conference is equal parts church revival, used car auction, and rap concert. I loved it."

From celebrity interviews to sporting news, *Sport Bloggers Live* uses its network of fanatical sports bloggers to help bring the news to your iPod. It's an inspirational example of citizen journalism: AOL members banding together to help report on the week's sporting news. The approach makes it feel fresher and a lot more down to earth. You won't find a nightly news reporter asking Tyson about reality TV.

The Cubicle Escape Pod

URL:	*http://www.cubicleescape.com/*
Frequency:	Biweekly
Duration:	30 to 50 minutes
Rating:	G

Podcasting about a new company is risky business. Jonathan Brown and Matt Thompson found out the hard way when they announced their company and product names on their show. In less than 24 hours someone had stolen the web address they planned to register. "There's no mercy rule in business" promptly became Jonathan's sixth business rule.

Jonathan and Matt intend to start a company that will sell a software product that currently doesn't exist.

On the journey, they plan to involve the listeners in decisions they make, information they learn, and in the end, the product that they release to the market. To get there they need knowledge from others, so they invite experienced guests on the show to talk about mission statements, branding, budgets, and the process of incorporating a company.

Each week is like an informal boardroom meeting that will enlighten and entertain anyone interested in the business world. Who knows, these two could end up as the next Bill Gates or Michael Dell, and we can say we knew them when they had their moms on the show.

Diary of a Shameless Self-Promoter

URL:	*http://www.heidimillerpresents.com/*
Frequency:	Weekly
Duration:	30
Rating:	G

Heidi Miller is a shameless self-promoter—it's even the name of her podcast. In her industry, it helps to be able to sell anything, including herself. Miller is a corporate presenter, someone who is hired to provide live trade show presentations, video narration, and presentation consulting. As she puts it, she is a human bug light and gets paid to talk.

In the podcast, Miller goes much further than just talking. She applies her own promotion principles on each show and speaks to guests about a range of networking topics.

Self-promotion and networking are difficult tasks. As one of Heidi's listeners suggested, most of us feel uncomfortable and visualize car salesmen in cheap suits when the topic is raised. However, these skills are incredibly important in business, and the show provides useful advice and takes the edge off the shamefulness. It won't make you feel comfortable self-promoting, but you'll be comfortable knowing others feel just the same. Or, maybe you'll feel like a human bug light.

Tips from the Top Floor

URL:	*http://www.tipsfromthetopfloor.com/*
Frequency:	Daily
Duration:	5 to 10 minutes
Rating:	G

They say that only mad dogs and Englishmen go out into the midday sun. Well, it seems that German photographers go out into midnight storms. At least Chris Marquardt does to take photos of lightning. While snapping the photos, he records a podcast, explaining how to get the perfect shot. His tips: get a nice composition, use a beanbag to steady the camera, and take as many photos as you can to capture the right moment.

Chris runs his own company, Top Floor Productions, based in Tübingen, Germany. It specializes in sound recording, photography, and web design—the perfect mix for a podcast about digital photography.

Almost every day Chris produces a nice and short show that consists of tips about digital photography. He aims the lessons at amateurs, but even the more experienced photographer can learn a thing or two from this very accessible show. All the tips are very simple but effective and would take years to learn without help.

While Germans are said to be staid and conservative, the stereotype doesn't fit Marquardt when he's sitting at his windowsill taking photos of storms, waiting to catch that one amazing photo.

Tom's Trucker Travel and Audio Podcast

URL:	*http://truckerphoto.blogspot.com/*
Frequency:	Several a week
Duration:	20 to 90 minutes
Rating:	G

Thomas R. Wiles sounds like any guy with a button-down shirt and tie. When he interviewed Rob Costlow, an up and coming solo pianist, you might have mistaken him for someone from the highbrow crowd who attends classical concertos. But Wiles looks as mean as hell. His tattoo-covered body, topped off with a skull tattoo on the back of his shaven head, says he's one mean dude.

But 50-year-old Trucker Tom, from Arkansas, isn't mean at all. In fact, he proves this time and again by spouting statements like, "If you row someone across a river, you get there yourself." He's a real nice bloke.

While in different parts of the country, Tom records his shows about computers, podcasting, movies, and trucks, and he reads and answers listeners' emails. Tom also shares moments from his trips (like his early morning arrival at Clifton, New Jersey, as everyone else is preparing for the day, slipping on work attire, and readying themselves for the daily commute). We hear all the bumps, squeaks, and thumps of the rig jolting in the background as he drives into the town and tells us about his trip.

Media Artist Secrets

URL:	*http://www.mediaartist.com/*
Frequency:	Several a week
Duration:	10 minutes
Rating:	G

The glamour of studio photography includes stunning models, makeup artists, fashion, swimwear, and the catwalk. Franklin McMahon knows all about it—he's been involved for years, and he draws from that experience in his show aimed at helping other artists build their career. In fact, McMahon is so hip he says he gave up swimwear photography because it was boring; it only allowed him to tell a limited story. (We really can't believe he thinks it's boring.)

McMahon's tips are gleaned from many sources, and he'll point listeners to books, tools, web sites, and magazines in a collage of examples to make a point. It's usually a short show, but it'll help inspire anyone who aims for success, and you too can progress to snubbing swimsuit photography in the name of art.

Fly with Me

URL:	*http://joepodcaster.libsyn.com/*
Frequency:	Occasional
Duration:	20 minutes
Rating:	G

Flying, whether you're a pilot or a flight attendant, is one of those careers everyone believes is glamorous. When pressed, these jetsetters will probably tell you the downside to the job: strange hours, tiring jetlag, and demanding customers. That doesn't stop Captain Joe d'Eon from making it sound as fun as we imagine it. Celebrities, pranks, jokes, and water canon salutes at retirement—it sounds like a blast to us.

In his show, Joe is a storyteller first and a pilot second. He treats us to tales, interviews, and snippets from air traffic control broadcasts. He'll collar fellow workers in the van on the way to the airport and quiz them about some of their favorite stories. Like the flight attendant who had to douse Richard Pryor with iced tea when he set his lap alight with a cigarette. Or the one about the two flight attendants who were picked to work on a flight dedicated to the Rat Pack: Frank Sinatra, Dean Martin, and Sammy Davis, Jr.

It's fantastic listening to what goes on at the front of the plane. Joe shows that the people on the other side of the door are human after all, and after listening to a few shows, you won't see the flight crew on your next trip in the same light. You'll have a new-found appreciation for their efforts and dedication. You might even be offering to copilot the plane because you know Captain d'Eon.

How to Do Stuff

URL:	*http://tmaffin.libsyn.com/*
Frequency:	When Tod gets a chance
Duration:	Less than 10 minutes
Rating:	G

According to *The Globe and Mail*, Tod Maffin is one of Canada's most influential futurists. He's a technology expert in business and media, and he founded an artificial intelligence company. Currently, he's a national technology columnist for CBC Television's *Canada Now*, as well as a national CBC radio host and producer. Anybody would wonder what gems of human wisdom the *How to Do Stuff* podcast brings to the world. Well, nothing that complex actually; it covers subjects like how to hammer a nail, how to clip a cat's nails, how to pour and drink the perfect Guinness, and how to drive in London.

Of course, Tod doesn't know the intricacies of all these common tasks, so he hunts down people who do.

The short show is professionally done, and it's easy to listen to and learn at least a little from the guests—even if they might be guessing all the answers.

DVD Talk Radio

URL:	*http://www.dvdtalkradio.com*
Frequency:	Weekly
Duration:	10 to 30 minutes
Rating:	G

Geoffrey Kleinman produces four different types of podcasts: reviews, recommendations, commentaries, and interviews. Some shows include all four and others a combination of these different genres. The informality of the show makes this 30-minute podcast a must-download for movie fans. Highlights of recent shows include a discussion with Zoe Bell, New Zealand's stuntwoman extraordinaire who doubled for Uma Thurman in *Kill Bill: Volume I* and *II*, telling how her arm got "mangled" in a stunt that went wrong; a talk to ghost hunters who use the Electronic Voice Phenomenon (EVP) to eavesdrop on spirits, a method explored in Michael Keaton's movie *White Noise*; and when Lloyd Kaufman, director of budget flick *The Toxic Avenger* and the how-to film *Make Your Own Damn Movie*, pretends his career goal was to be interviewed on *DVD Talk Radio*.

If you hunger after quality information about up-and-coming DVDs, whether blockbusters like *The Incredibles* or B-movies like *Mako*, then *DVD Talk Radio* will help provide your weekly fix.

Coverville

URL:	*http://www.coverville.com*
Frequency:	Roughly twice a week
Duration:	30 minutes
Rating:	G

What does any self-respecting music geek do with his collection of 3,500 compact disks and vinyl records? If you're Brian Ibbott, you produce a podcast dedicated to cover songs, music performed by artists other than the originals.

Each week he compiles cover songs from his collection and records a podcast in the basement of his house in Denver, Colorado. Listeners or independent bands often make recommendations, and from these, Ibbott produces several 30-minute shows. They include a wide range of cover tunes, from a version of the Genesis hit "I Can't Dance," performed by Gnemesis on two ukuleles and an electric kazoo, to an electronic version of Nirvana's hit "Smells Like Teen Spirit" by Moog Cookbook, all the way to a punk version of the recent Kalis hit "Milkshake," performed by Chance.

Coverville is like a radio show, featuring songs punctuated by Ibbott's comments and stories. The show sometimes surfs on a theme for its entire 30 minutes; this could be an artist, tunes that are somehow related, or a special event. Examples include a Burt Bacharach special, a show dedicated to The Squeeze, a Double Double Cover Cover show—where artists "wedge" a song into another cover—and a Valentine's Day special.

The Treatment

URL: *http://www.kcrw.com/show/tt*

Frequency: Weekly, aired on Wednesdays

Duration: 30 minutes

Rating: G

KCRW is a community radio station based at Santa Monica College in California. One of their most popular shows is *The Treatment*—named after a film industry term for an overview of a screenplay—broadcast once a week across the United States on a variety of stations as part of National Public Radio (NPR).

With guests such as director Christopher Nolan and actor Sean Penn, the show doesn't need much publicity to attract an audience. The combination of such well-known actors and directors, and Mitchell's deep understanding of film, makes this a well-balanced show. Mitchell's questions delve deep into the psyche of the film he's discussing, yet he pulls no punches and sometimes asks delicate questions. His confidence about the cinema is evident, but Mitchell isn't all business all the time. He couldn't stop laughing when he interviewed the famous Eric Idle, former member of Monty Python.

The Treatment became available as a podcast in May 2005, after KCRW realized that it could reach more listeners through the Internet.

Insomnia Radio

URL:	*http://www.insomniaradio.net*
Frequency:	Weekly
Duration:	30 minutes
Rating:	PG (some lyrics may offend)

After hosting his *Insomnia Radio* podcast for six months, Jason Evangelho, a butcher and computer technician in California, got a call from one of the largest radio broadcasting companies in the United States. The conglomerate owns some 180 radio stations in 22 states, and they wanted to add *Insomnia Radio* to their lineup. Jason Evangelho said no. He wanted to be free to play the music he chose, and he felt that the network would restrict his choices. His passion for independent music would lead him to feel guilty for allowing a corporation to dictate what he played on his show. Right on, Jason!

Jason has so much enthusiasm for independent music that it overflows from his weekly podcast. Originally called *Hardcore Insomnia Radio*, he renamed it to be clear he didn't focus on punk, hardcore, or metal. In fact, he plays a wide variety of music, from all genres, by independent bands that have never signed with major labels.

Most of the music on *Insomnia Radio* is by American bands, but Jason selects the occasional tracks from bands from other parts of the world, such as Canada and Portugal. While the focus is usually on indie rock, he sporadically goes for a change of pace, playing pop, electronica, or grunge tunes.

Slice of SciFi

URL:	*http://www.sliceofscifi.com*
Frequency:	Weekly
Duration:	45 minutes
Rating:	G

It's not surprising that *Slice of SciFi*, a podcast dedicated to science fiction news, spent its first few shows talking about the campaign to save the TV show *Star Trek Enterprise*. However, since the demise of *Enterprise*, the over-the-top hosts, Michael and Evo, have found plenty more to talk about in the universe of sci-fi. They've featured television shows such as *Doctor Who*, *Firefly*, and *Battlestar Galactica*, as well as movies, such as the pre-screening for *Batman Begins*.

Most of the show focuses on science fiction news, including movies, television, and literature. It's all-important news, like the eight-year-old student who was suspended for replacing the Pledge of Allegiance with his own pledge to the flag of the United Federation of Planets. They also hunt down guests, including Eugene Roddenberry, Jr. (son of Gene, the creator of the Star Trek franchise) to discuss the work he's doing on the Trek Nation documentary.

You don't need to don Vulcan ears or wield a lightsaber to enjoy the gang talk about science fiction. If you enjoy the genre on TV, on the silver screen, or in books and magazines, then Slice is the right channel to tune to. Beam us up, Michael and Evo.

Parking in the Bitterman Circle and Modern Roadie

URL:	*http://www.bittermancircle.com*
	http://www.modernroadie.com
Frequency:	Irregular; several recordings per month
Duration:	10 to 30 minutes
Rating:	PG

While at a gig, waiting for your favorite band to enter, a lone dark figure struts onto the stage. The crowd roars in appreciation for just a moment and then hushes in disappointment. As Aron Michalski puts it, "I'm just a guy who sets stuff up and makes sure it works so other people can use it. I'm a roadie."

After listening to Aron, a drum and bass tech for the band Weezer, you'll learn to appreciate roadies and wonder how they can still hear anything. Aron, who has worked in the music business for 20 years for artists including Bruce Springsteen, provides a podcast containing such content as monologues of written notes from 2002 and recordings of the team doing a festival changeover at the Nova Rock Festival in Nickelsdorf, Austria.

Across the Atlantic in England, Paul Eastman, a keyboard tech for the Manchester pop trio Doves, takes listeners along with the band as they travel to gigs around the world.

Both roadies provide a diverse range of material to listen to. There are no drugs or sex, but plenty of rock and roll and backstage banter.

TV Guide Talk

URL:	*http://www.tvguide.com/news/podcast/*
Frequency:	Weekly
Duration:	30 minutes
Rating:	G

TV addicts, rejoice! This podcast provides the latest news straight from the staff at TVGuide.com. These guys live television, literally; they've probably got antennas hooked up to their brains to get a constant feed. Michael Ausiello, the Guide's News Director and a popular TV show critic, is so much a part of this world that he once made a guest appearance in one of his favorite shows, the *Gilmore Girls*.

For each podcast, the TVGuide.com staff, Lauren Ruotolo, Maitland McDonagh, and Daniel Manu, join Ausiello to give an excellent breadth of coverage. They start the show with news and sweep into segments such as Ask Ausiello, which answers questions from readers and listeners, and McDonagh's movie reviews.

Probably the most fun in the show is Ausiello's playful, catty comments, such as his feelings about Tom Cruise's antics on *Oprah*. The crew also discuss their online magazine articles and critiques of new shows or series that have hit the tube.

Only those immersed in the entertainment industry could match this show for content, and only The Fab Five, from *Queer Eye for the Straight Guy*, can match their cattiness.

Rock and Roll Geek Show

URL:	Podcast: *http://rockandrollgeek.podshow.com/*
	Stream: *http://www.live365.com/stations/adamc1999*
Frequency:	Weekly with other shows interspersed
Duration:	30
Rating:	PG

The men's bathroom at a rock and roll gig is the last place you want to go on a podcast soundseeing tour. But that doesn't stop Michael Butler from taking you there when he goes to a bar for a concert by the band Tsar.

These raw moments are what make the *Rock and Roll Geek Show* appealing to rock fans. Butler, a bass player for a Californian rock band, says he likes music that is "hard rock with a catchy chorus." So that's what he selects for the show.

Butler plays songs from independent rock bands, interviews guests, and has a special segment he calls "Listening with Butler." In this part of the show, as he plays a song he's never heard he gives his opinion, like, "that first line is lame" or "it's pretty catchy pop punk."

The show's growing popularity has allowed Butler to interview some real rock legends, such as Christina Amphlette, singer for the Divinyls. Surely, that's any head-banger's dream.

Radio Clash

URL:	*http://www.mutantpop.net/radioclash/*
Frequency:	Weekly
Duration:	40 to 50 minutes
Rating:	PG

Tim Baker, a London-based designer, grew tired of regular radio and decided to dabble in mashing-up tunes after hearing a mix of TLC versus Human League in a music store. After learning about podcasting, Tim realized that it was a great way of sharing the tunes he created or heard, and started *Radio Clash*. The name is a pun on the concept of mashing or clashing music, and is also the title of a song by The Clash.

In the weekly 40- to 50-minute show, Baker plays and discusses mashup music from around the world. He intersperses this with rants about personal topics, discussions of music, and occasional unrelated interests, like the new British *Doctor Who* television series. When he's got a guest on the show, the conversation can range from mashups and techniques to the commonality of the use of the word *douche* in songs.

Mashups aren't for everyone: some a capellas, tracks with just vocals, sound downright strange when mixed with a tune, and sometimes they sound off-key. However, they do make an excellent party game of "guess the artists."

The Gadget Show

URL:	*http://www.thepodcastnetwork.com/gadget/*
Frequency:	Weekly
Duration:	20 to 40 minutes
Rating:	PG

The Gadget Show doesn't just talk about the dinky toys that we carry in our pockets. The show's host has interviewed the lead ground controller for NASA's shuttle missions, a movie buff who had a purpose-built home theatre with a floating floor and a suspended ceiling, and über-geeks famous for their exploits on the Internet.

The host of the show (and one of the authors of this book), Richard Giles, works for Sun Microsystems and has played with some of the big toys, such as a computer larger than your average fridge.

Richard doesn't just chat about serious big-iron computers; he brings a laidback Australian approach to the show. His interviews are very informal, providing a breather from scripted talk radio.

He once called the phone booth in front of Grumman's Chinese Theatre in Los Angeles. Fans had lined up for weeks in the cold to see the premier of *Star Wars: Episode III— Revenge of the Sith*. As chance would have it, there was an Aussie in the line who provided the views from the queue.

If you're keen on gizmos and hanker for a casual show about technology and the people immersed in it, then tuning in down under is the way to go.

Index

We'd like to hear your suggestions for improving our indexes. Send email to
index@oreilly.com.